Vanderlei Farias De Lima
Claudio Nahum Alves
Paola Souto Campos

Community management of pirarucu Arapaima spp. in the Mamirauá RDS

Vanderlei Farias De Lima
Claudio Nahum Alves
Paola Souto Campos

Community management of pirarucu Arapaima spp. in the Mamirauá RDS

An analysis of the Jarauá Sector

ScienciaScripts

Cover image: www.ingimage.com

This book is a translation from the original published under ISBN 978-613-9-61014-3.

Publisher:
Sciencia Scripts
is a trademark of
Dodo Books Indian Ocean Ltd. and OmniScriptum S.R.L publishing group

120 High Road, East Finchley, London, N2 9ED, United Kingdom
Str. Armeneasca 28/1, office 1, Chisinau MD-2012, Republic of Moldova, Europe
Printed at: see last page
ISBN: 978-620-7-50088-8

Table of contents:

CHAPTER I

INTRODUCTION

Fishing in the Amazon region is considered one of the most traditional and important extractive activities for its population (BATISTA *et al.* 2012; GARCEZ; BOTERO; FABRÉ, 2012). The main source of animal protein and income for riverside families, fishing in this region differs from other Brazilian regions due to the diversity of species that are exploited, the quantity of fish caught and the dependence that rural communities have on this resource (ARAÚJO; FREITAS, 2009; ISAAC *et al.*, 2012).

According to Ruffino (2014), the Amazon has more than 3,000 species of fish, with commercial and subsistence exploitation concentrated on approximately 200 of them, especially tambaqui (*Colossoma macropomum),* pirarucu (*Arapaima gigas*), jaraquis (*Semaprochilodus* spp.), matrinxa (*Brycon amazonicus*), curimata (*Prochilodus nigricans*), pacus (*Mylossoma* spp. and *Myleus spp.),* sardines (*Triportheus spp.),* piramutaba (*Brachyplatystoma vaillantii),* surubim (*Pseudoplatystoma fasciatum*), douradas (*Brachyplatystoma spp.*) and tucunarés (*Cichla spp.)*.

Fishing production in the Amazon region is estimated at around 217,000 tons per year, one of the highest rates in the world, equivalent to between 20 and 25% of all fishing income in Brazil, both marine and freshwater (COSTA *et al.*, 2013). The average per capita consumption of fish in the Amazon is 369g/day (135 kg/year), while in Brazil it is 20g/day (7 kg/year) and in the world 16 kg/year, i.e. considering the consumption of 12 kg/year recommended by the World Health Organization (WHO), it is clear that in the Amazon fish is consumed more than ten times the values indicated (SANTOS; FERREIRA; VAL, 2010).

Although the availability of fish is abundant in this region, from the 1960s onwards, the decline of the main floodplain economic activities, such as the exploitation of rubber and jute, meant that factors such as the growth of large urban centers, the introduction of new fishing technologies and tax incentives for the sector led to the intensification of fishing in the Amazon (PETRERE *et al.*, 2007;

RUFFINO *et al.*, 2012),

especially in the concentration of catches of species of greater commercial value such as *Arapaima* spp. commonly known as pirarucu (BARTHEM, R. B.; GOULDING, 2007; CASTRO; MCGRATH, 2010; JÚNIOR; ALMEIDA, 2006).

Regionally known as the "cod of the Amazon" due to its tasty meat, good quality and high nutritional value, the pirarucu has been one of the main products in the diet of the local Amazonian population since pre-colonial times, Compared to dried meat from southern Brazil and cod (*Gadus morhua*) from Europe and North America, in the 19th century the process of salting this fish was so important that it was characterized by a time of year called "salting time", which took place between September and October (HRBEK; CROSSA; FARIAS, 2007; SANTOS, 2014).

According to Viana *et al.* (2007), due to overexploitation, a significant decrease in the catch of this species began to be noticed in the 1970s, and it became commercially extinct near the large urban centers of the Amazon. In the same decade, the pirarucu became part of Appendix II of the Convention on International Trade in Endangered Species of Wild Fauna and Flora (CITES), which lists species that are not necessarily threatened with extinction, but could become so if international trade is not controlled (CASTELLO; STEWART, 2010; MUELLER; GREEN, 2006).

Since then, more restrictive exploitation measures have been taken by the Brazilian Institute for the Environment and Renewable Natural Resources (IBAMA), which in 1989 established a minimum catch size of 150 cm in total length, in 1990 the reproductive "closed" period from December 1st to May 31st and, in 1996, the prohibition of any type of capture and commercialization of pirarucu in the state of Amazonas, being allowed only in management areas or from cultivation (Ordinance 8/96) (AMARAL *et al...* 2011; ARANTES 2009a), 2011; ARANTES, 2009a).

This ban has affected countless riverside families in the state, due to the importance of its fishing in the income of these populations (LIMA, 2010; PERALTA *et al.,* 2009). However, with the aim of reversing the situation and subsidizing its management, research was carried out into its biology, ecology and fishing, and the first community pirarucu management project in the state of Amazonas was submitted to and subsequently approved by IBAMA, starting in 1999 in the Mamirauá Sustainable Development Reserve (RDSM) (AMARAL; ALMEIDA, 2013; COUTINHO, 2008).

Nineteen years after its implementation in the Mamirauá RSD, community pirarucu management is making progress and is being replicated in various municipalities in Amazonas such as Anori, Barcelos, Beruri, Carauari, Coari, Codajás, Fonte Boa, Japurá, Juruá, Jutai, Maraa, Uarini, Tapauá and Tonantins, and other states such as Acre and Pará, as well as in other Pan-Amazon countries such as Peru, Colombia and British Guiana (BESSA; LIMA, 2010; GONQALVES, 2013).

Considering the economic, social and cultural importance of the pirarucu fishery for riverside communities, its management is essential not only in guaranteeing the livelihoods of these populations, but also in the growth of their stocks, because although it is a resistant species, its biological and ecological characteristics such as the onset of gonodal maturation, which only occurs in the fourth or fifth year of life at 1.60 to 1,70 meters in total length and about 40 to 50 kg in weight, low fecundity, the need to climb to the surface at intervals of 5 to 15 minutes to perform obligatory aerial respiration and the period of protection for the offspring, making it easy prey for fishermen, in addition to the natural predation of the fry by piranhas after parental capture, mean that the reproductive success of the species is diminished (CASTELLO; STEWART; ARANTES, 2011; GODINHO *et al.*, 2005; LIMA; BATISTA, 2012; LOPES *et al.*, 2013).

This work therefore sets out to carry out an economic, social and environmental analysis of the community management of pirarucu (*Arapaima* spp.) practiced in the Mamirauá Sustainable Development Reserve, in the state of Amazonas, between 1999 and 2015. The analysis is based on a specific sector of the reserve's focal area: the Jarauá Sector, since this sector was the first to develop this type of fishing management in the region.

The data used comes from a secondary source, collected from the annual reports on managed pirarucu fishing in the Mamirauá RDS, available from the IBAMA Amazonas Superintendence and previously authorized by the authority's Fauna Center.

CHAPTER II

Species description

The pirarucu (Figure 1) belongs to the genus *Arapaima*, family Arapaimidae and order Osteoglossiformes. It is considered to be the largest freshwater scaled fish in the world, measuring approximately 3 meters in total length and weighing more than 200 kg, as well as being considered the largest "living fossil" on Earth, belonging to one of the oldest groups of living teleosts that emerged during the Jurassic period (165 million years ago), possibly in Pangea (ARANTES, 2009b; HRBEK *et al.,* 2005).

Figure 1: Male *Arapaima gigas* with a total length of 153 cm caught on November 22, 2006, in the Panema sandbank, Jarauá Sector, in the Mamirauá RDS, Amazonas. Photo: Donald J. Stewart.
Source: Arantes, 2009b.

This species has a small, flat head and a long, cylindrical body covered in scales with a normal coloration that varies between brown, tan and greenish, depending on the water it is found in. Its scales are cycloid and large, arranged over its body like a mosaic, the edge of which is colored red mainly on the tail, belly and flanks, being more predominant in males during the reproductive period (CASTELLO; STEWART; ARANTES, 2011).

This species of *Arapaima*, known in Brazil as "pirarucu" as an adult and "bodeco" as a juvenile, despite being endemic to the Brazilian Amazon basin due to its preference for the lacustrine environments of its floodplains (Figure 2), has records in Peru, Colombia, Bolivia, Ecuador and Venezuela, where it is known as "paiche" and in the Guianas as "warapaima" (LOPES; QUEIROZ, 2010). Its popular name "pirarucu" originated in indigenous culture, meaning *pirá* (fish) and *urukú* (red), due to the red color of its tail, belly and flank scales, a characteristic color of the annatto seed (*Bixa orellana L.*) (SANTOS; FERREIRA; ZUANON, 2006).

Figure 2: Specimens of *Arapaima* spp.
Source: Ferraris, 2003.

Considered the most famous and emblematic fish of the Amazon ichthyofauna, the *Arapaima gigas* was considered the only species of pirarucu in the world more than 140 years ago, but recent studies show the existence of four other species of *Arapaima, the A. arapaima,* the A. *mapae* and the *A. agassizii. agassizii*, already classified by scientists Cuvier and Valenciennes in 1847, but disregarded by Günther in 1868, and finally *A. leptosoma*, described in 2013 by ichthyologist Donald Stewart, caught in 2001 at the confluence of the Purus and Solimoes rivers in Amazonas, bringing the number of species to five (Figure 3) (CASTELLO; STEWART; ARANTES, 2013; STEWART, 2013a; STEWART, 2013b).

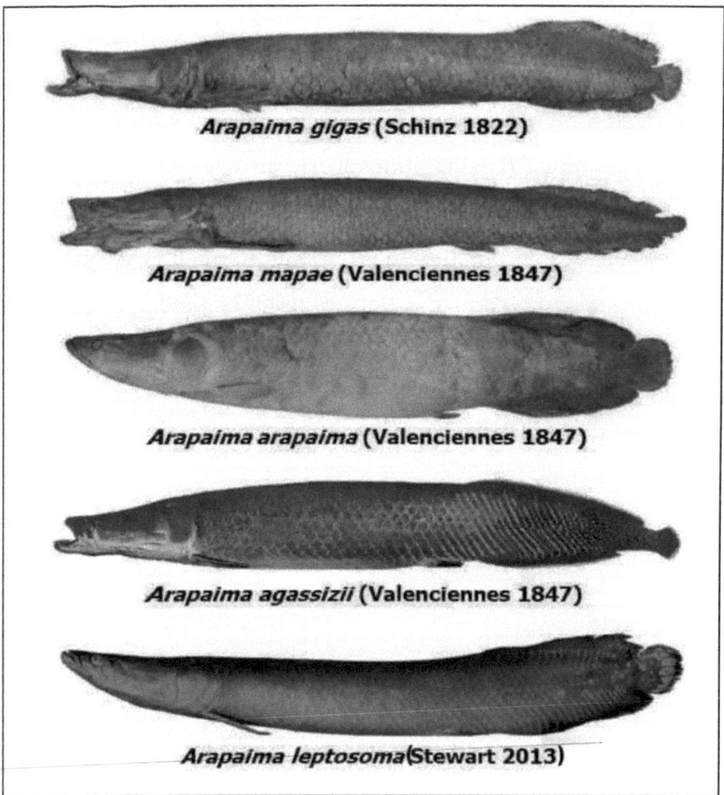

Figure 3: Different species of pirarucu.
Source: Stewart, 2013a.

It has a carnivorous/piscivorous eating habit, consuming preferably caracid and loricariid fish, but also shrimps, crabs and insects. Its tongue is made up of a rough, bony plate that has the function of holding and crushing its prey. It is a predator at the top of the food chain, is territorial, invests care in the survival and longevity of its offspring and has a sedentary behavior (CASTELLO, 2007).

The reproduction of the species takes place in stages and is closely related to the climatic factors characteristic of the Amazon, such as the variation in rainfall and temperatures throughout the year. The high temperatures combined with the rainfall are the main stimulus for the reproductive process, which involves mating, spawning, incubation, hatching of the eggs and caring for the larvae and fry. It is known that a female pirarucu is capable of producing around 11,000 fry per spawning, which is low compared to other fish species (ARANTES; GARCEZ; CASTELLO, 2006; LIMA;

BATISTA, 2012).

Throughout its life, the pirarucu has two systems for obtaining oxygen: the gills for gill breathing and a highly vascularized swim bladder, which functions as a lung for air breathing. They must breathe atmospheric air every 5 to 15 minutes, as their gills are not enough to capture all the oxygen they need from the water. During aerial breathing, the fish releases carbon dioxide (CO_2) and immediately absorbs oxygen (O_2) (Figure 4). When it wants to go down, it empties its swim bladder and when it goes up, it fills it with the air it takes from the water. It is with this upward movement that fishermen are able to estimate how many fish there are in an area or catch them (ARANTES; CASTELLO; GARCEZ, 2007; CASTELLO, 2007).

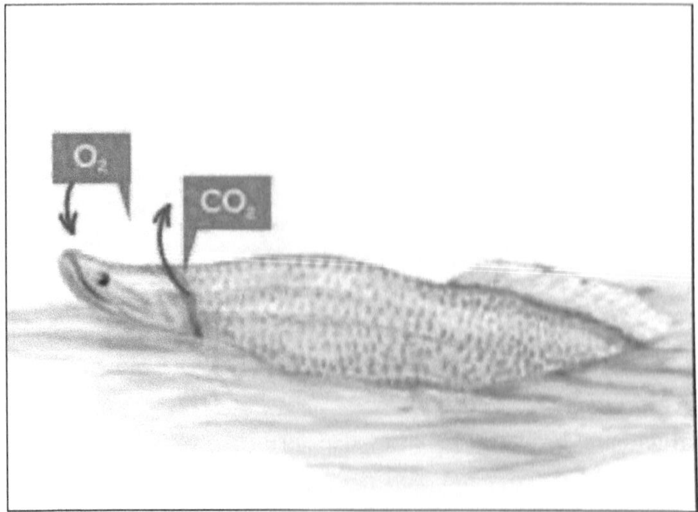

Figure 4: Pirarucu breathing system.
Source: Castello, 2007.

The pirarucu's bladder also functions as a resonance box that amplifies the sound picked up through the fish's lateral line (Figure 5), used to detect movements and vibrations in the water, as well as aiding locomotion in environments with poor visibility and obstacles (CASTELLO; STEWART; ARANTES, 2011).

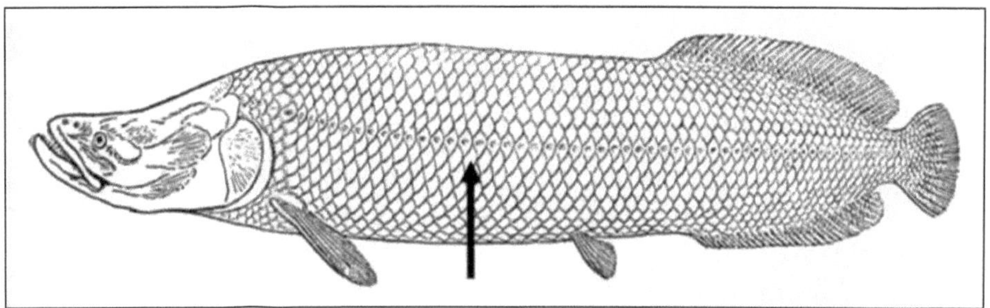

Figure 5: Pirarucu's lateral line used to capture sound.
Source: Jornal da USP, 2016.

As for the methods used to catch pirarucu, there are restrictions on the use of fishing tools or methods established by IBAMA Ordinance 008/96 for the Amazon, Araguaia and Tocantins river basins, which prohibit trawls, cast nets and electric nets, corrals or traps of the tapagem type and methods that use batipao, toxins or explosives. Nets are limited to lengths and modes of use, with the harpoon being the main tool used by fishermen during fishing (Figure 6) (IBAMA, 1996).

Figure 6: Method of fishing for pirarucu using arpao.
Source: Mamirauá Institute, 2014.

Pirarucu meat is naturally pink in color and devoid of bones, making it highly prized. It is considered to be a food rich in protein, outperforming the meat of salmon, sardines and beef when subjected to the same treatment. It is mainly sold in the Amazon region in the form of chaturo (Figure 7), fresh mantas (Figure 8) and salted mantas (Figure 9) (ARARIPE, 2008).

Figure 7: Pirarucu in the shape of a cigar. **Source:** Araripe, 2008.

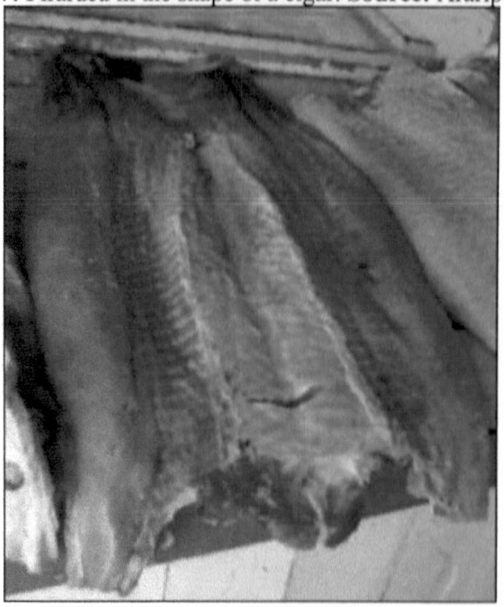

Figure 8: Fresh pirarucu blanket.
Source: Araripe, 2008.

Figure 9: Salted pirarucu blanket.
Source: Araripe, 2008.

In addition to the meat, the processing of pirarucu also offers some by-products such as the scales (Figure 10), the tongue (Figure 11), the leather (Figure 12), the head, the viscera and the fins, which can be reused to add value to the production and provide the creation of new products (DE CARVALHO PAIVA *et al.*, 2015).

Figure 10: Pirarucu scales.
Source: Flickr, 2006.

Figure 11: Pirarucu tongue.
Source: Forest objects, 2012.

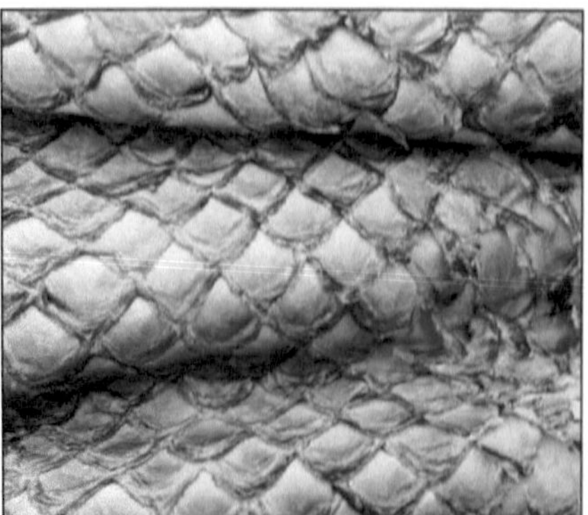

Figure 12: Pirarucu leather. **Source:** Péltica, 2017.

The scales are used to make handicrafts such as curtains, key rings and lampshades, and to produce bio-jewelry such as necklaces, earrings and bracelets. The pirarucu's bony tongue is used in the Amazon by traditional peoples as a grater for guaraná and Brazil nuts. The leather has great potential for making bags (Figure 13), shoes (Figure 14), belts and clothing (WWF, 2011).

Figure 13: Pirarucu leather pouch.
Source: Taia, 2017.

Figure 14: Pirarucu leather shoe. **Source:** Silverado boots, 2017.

According to studies by the National Institute for Amazonian Research (INPA), pirarucu leather has advantages over bovine leather: it sustains more tension, since its fibers are interlocked, while bovine fibers are arranged in parallel. Tanned, dyed and well-finished by specialists, pirarucu leather has won the market in France, Italy, the United States and Canada (NÓBREGA, 2015).

Community pirarucu management

Before management, pirarucu fishing in the Reserve took place all year round.

It used to be sold in the port of the communities by each family and in small quantities to middlemen (regatóes), who received the fish in payment for goods or benefits made by them to the communities. Under the management system (Figure 15), pirarucu fishing began to take place during the dry season (between October and November), respecting the "closed season", facilitating the logistics of fishing and selling and also the control of the inspection body. This fishing began to be done in groups and sold through community associations (AMARAL, 2009).

Figure 15: Flowchart of pirarucu management activities in Mamirauá throughout the year. **Source:** Amaral, 2009.

Management in the Mamirauá Reserve makes use of lakes and other environments in the floodplain ecosystem such as canos, paranás and ressacas, which are suitable for managing species such as pirarucu (Figure 16). These environments are currently grouped into three socio-environmental contexts: lakes for the exclusive use of riverine fishermen; lakes for the exclusive use of urban fishermen; and lakes for shared use by urban and riverine fishermen. Following the implementation of the counting method, the rotation of lakes ceased to be the basic principle for management and other fishing criteria began to be considered, such as the quantity of fish and ease of access to the managed environments (AMARAL; QUEIROZ, 2011; CASTELLO,

2004).

Figure 16: Fishermen fishing for pirarucu in a Mamirauá lake.
Source: Mamirauá Institute, 2017.

According to Castello (2004), the counting methodology estimates the abundance of pirarucu in a given area based on the fish's aerial breathing. Counting also distinguishes pirarucu individuals by size class, with those smaller than 1.5 meters in total length being considered juveniles and those larger than 1.5 meters in total length being considered adults.

This methodology was validated by a study carried out in the Mamirauá RDS between 1999 and 2000, which compared the counts of a group of 20 fishermen with abundance estimates made by marking and recapturing individuals. The study showed that these counts varied by only around 10% from the true abundance value (CASTELLO, 2004).

To count pirarucu there is a training course on the application of the fish counting method, which combines scientific knowledge with the traditional knowledge of fishermen to carry out a population survey of pirarucu in bodies of water in natural environments, and consists of three stages: theory, practice and evaluation (Figure 17) (SILVA; GONÇALVES; MARINHO, 2013).

Figure 17: Representation of the training steps in the use of the pirarucu counting method.
Source: Silva, Gonçalves, Marinho, 2013.

CHAPTER III

RDS Mamirauá

The Mamirauá Sustainable Development Reserve (RDSM) (Figure 18) is located in the state of Amazonas near the city of Tefé, about 600 km west of Manaus, at coordinates 03°09'35"S 64°47'37"W, 01°50'05"S 65°42'19"W and 02°32'50"S 67°22'08"W.

Figure 18: Location map of the Mamirauá RDS.
Source: IDSM, 2014.

It comprises an area of 1,124,000 ha of floodplains divided into two parts: the focal (main) area with 260,000 ha, and the subsidiary (complementary) area with 864,000 ha. It is delimited by the Solimoes, Japurá and Auati-Paraná rivers in the middle Solimoes river and falls within the territory of the municipalities of Uarini,

Fonte Boa, Japurá, Tonantins and Maraa, with an area of influence in other municipalities such as Jutaí, Alvaraes and Tefé (IDSM, 2014).

Initially established as an Ecological Reserve in 1990, six years after its creation it was re-categorized as a Sustainable Development Reserve, as it reconciled the human development of the local populations with the conservation of biodiversity in a sustainable way, and was the first of its kind to be created in the country (QUEIROZ; PERALTA, 2006).

Currently, the Mamirauá RSD is part of the Central Amazon Ecological Corridor and is part of the Amazon Biosphere Reserve, as well as being one of the Brazilian reserves recognized by the UN Ramsar Convention, which protects wetland environments of great international importance and is also considered a World Natural Heritage Site by UNESCO (United Nations Educational, Scientific and Cultural Organization) (Figure 19) (SACRAMENTO; SCHMITZ, 2013).

Figure 19: Aerial view of the Mamirauá RDS, highlighting the Uacari lodge.
Source: Mamirauá Institute, 2017.

According to data from the last demographic census of the reserve, there are 1,873 households, with 10,867 people, including residents and users, who are distributed in the focal and subsidiary areas of the reserve. The main economic activities carried out by these people are artisanal and commercial fishing, family farming, logging and community-based tourism (IDSM, 2014).

Jarauá Sector

The Jarauá Sector (Figure 20), coordinates (65°1'16.595"W 2°47'55.299"S), is located in the focal (main) area of the Mamirauá Sustainable Development Reserve (RDSM).

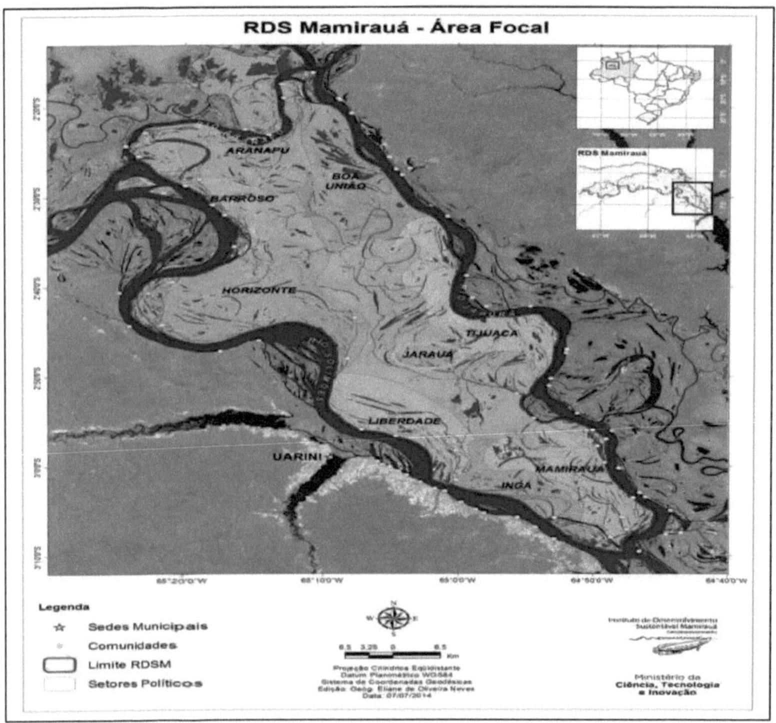

Figure 20: Location map of the Jarauá Sector in the Mamirauá RDS focal area.
Source: IDSM, 2014.

The Sectors are human groupings organized into communities that share the management and use of the Reserve's natural resources and were delimited taking into account the socio-economic, hydrological and ecological aspects of the site (IDSM, 2014).

The Jarauá Sector was the pioneer in 1999 in the pirarucu management experiment due to the fact that it houses the largest system of lakes in the focal area of the Reserve, due to the great fishing potential, the organization of the communities and the active participation of their representatives in the process of implementing the Mamirauá RDS (VIANA *et al.*, 2007).

This sector has the largest area (563 km^2) and the largest number of lakes (369), and is made up of four communities: Sao Raimundo do Jarauá (Figure 21), Nova Colómbia, Novo Pirapucu and Manacabi, which are home to a population of approximately 500 people (IDSM, 2014).

Figure 21: View of the Sao Raimundo do Jarauá community - Jarauá Sector, Mamirauá RDS. **Source:** Mamirauá Institute, 2017.

CHAPTER IV

Fishing production

The total production of pirarucu in the Jaraua Sector over 16 years of management amounted to 682,028 kg, which showed considerable growth from 1999, when 3,000 kg were produced, to 2015, with a production of 67,422 kg (Figure 22).

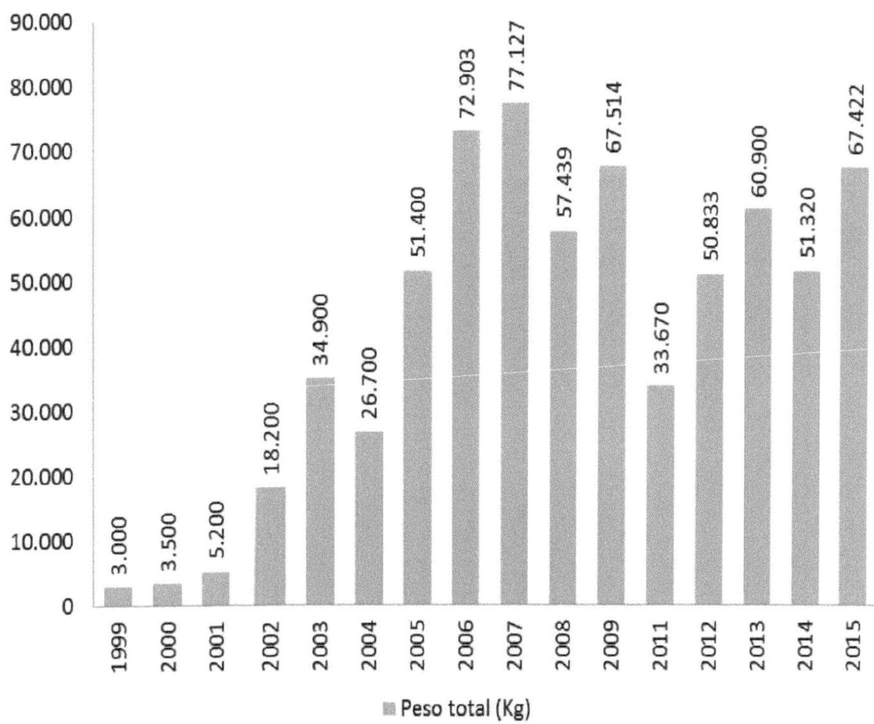

Figure 22: Evolution of production (kg) of pirarucu managed by the Jarauá Sector.
Source: Mamirauá Institute.

The best management performance in the Sector was in 2007, when production reached 77,127 kg. The lowest performance was observed in the first year of management, in 1999, with 3,000 kg, but this Sector shows good production results when compared to the last year analyzed (Figure 22).

It can be seen that despite the growth in production, there have been declines over the years. These declines can be explained by the number of lakes fished each year, the variation in the number of fish caught and the size of the individuals caught.

The production of the Jarauá Sector stands out due to the fact that this Sector concentrates the largest number of environments, 369 lakes. This means that the species is able to develop over a longer period of time and reach greater biomass, which will consequently result in greater production, even if the number of individuals caught is lower.

Between 1999 and 2015, managers caught an average of 90% of the authorized quota (Figure 23). This demonstrates the good performance of fishing groups in productive organization and confirms the growing availability of the resource in managed areas. Fishing quotas are determined on the basis of stock surveys. As pirarucu densities increase, production tends to grow.

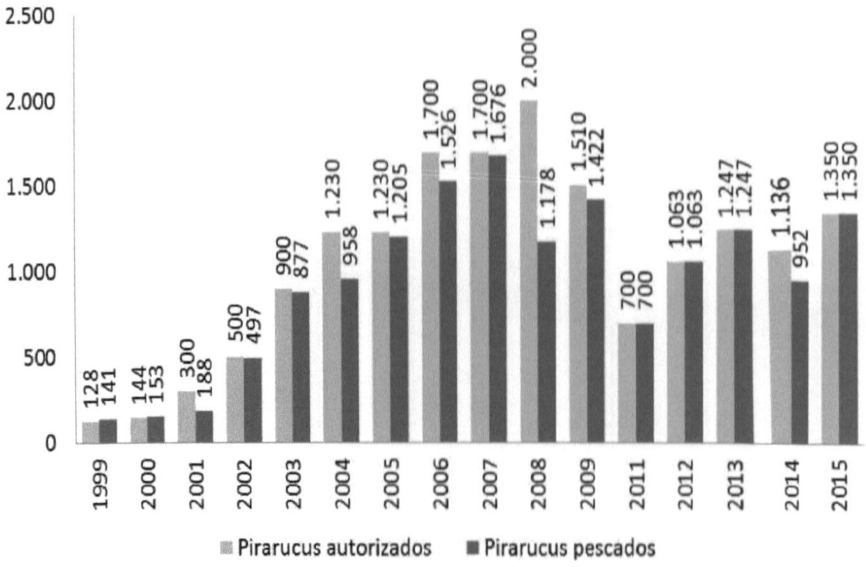

Figure 23 Relationship between authorized fishing and actual pirarucu fishing in the Sector.
Source: Mamirauá Institute.

The fishing quotas requested annually by the management groups from IBAMA represent around 15% to 30% of the number of adult pirarucus counted (fish over 1.5 meters) in the management lakes. The relationship between the authorized quota and the number of pirarucus caught by the fishermen provides an indicator of fishing effectiveness, since this relationship reflects both the availability of the resource, which complements the counts, and the organization of the fishermen to produce the fish (AMARAL; ALMEIDA, 2013).

Socio-economic performance

The number of fishermen benefiting over the whole period was accompanied by more fishermen joining the management program (Figure 24).

On average, around 6 fishermen have joined management each year, making for a 223% increase in the number of people involved from 1999 to 2015. As a result, more families have benefited from the income derived from management.

Figure 24 Number of fishermen benefiting from pirarucu management.
Source: Mamirauá Institute.

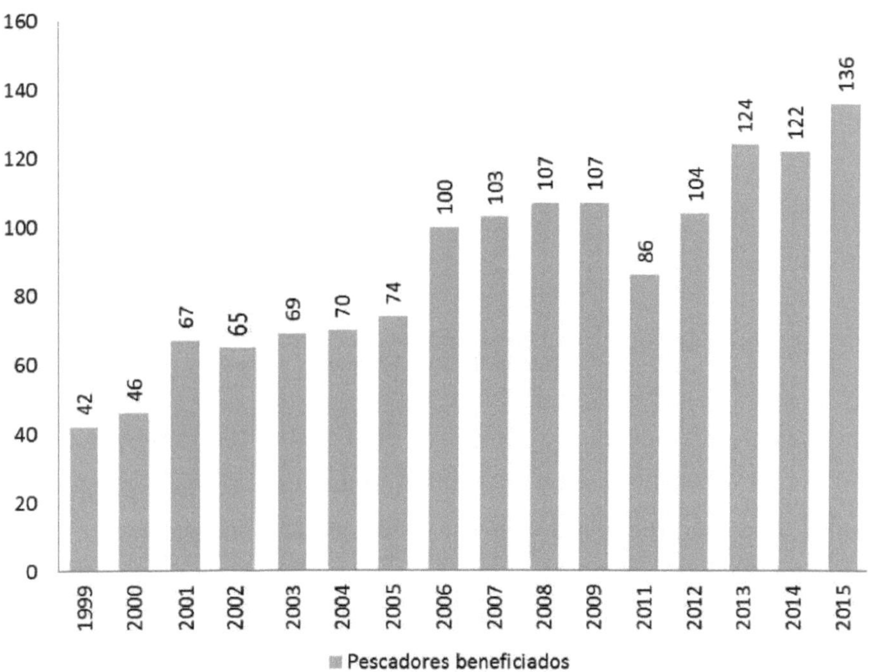

In terms of total gross revenue, there was an increase of 2,709% between 1999 and 2015 (Figure 25). However, possible problems in the marketing of production can generate economic losses, as well as discouraging fishermen and weakening participatory management as a whole.

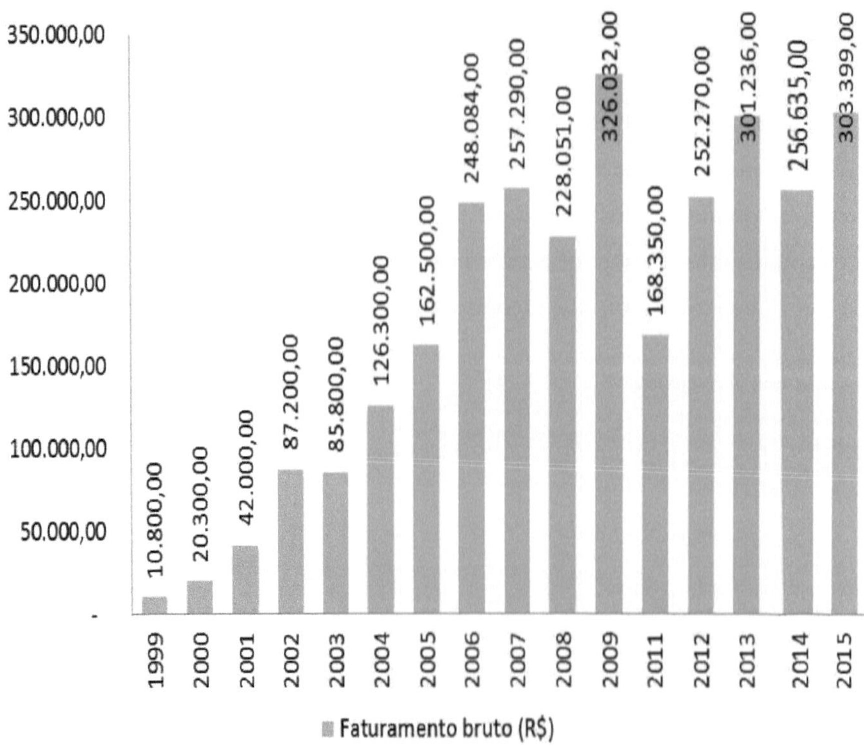

Figure 25 Gross revenue (R$) from the production of managed pirarucu in Jarauá.
Source: Mamirauá Institute.

The average gross revenue per fisherman, on the other hand, showed an upward trend of around 767%, followed by small falls over the years, probably due to the large number of fishermen who joined the management and the stabilization of the prices paid for the product (Figure 26).

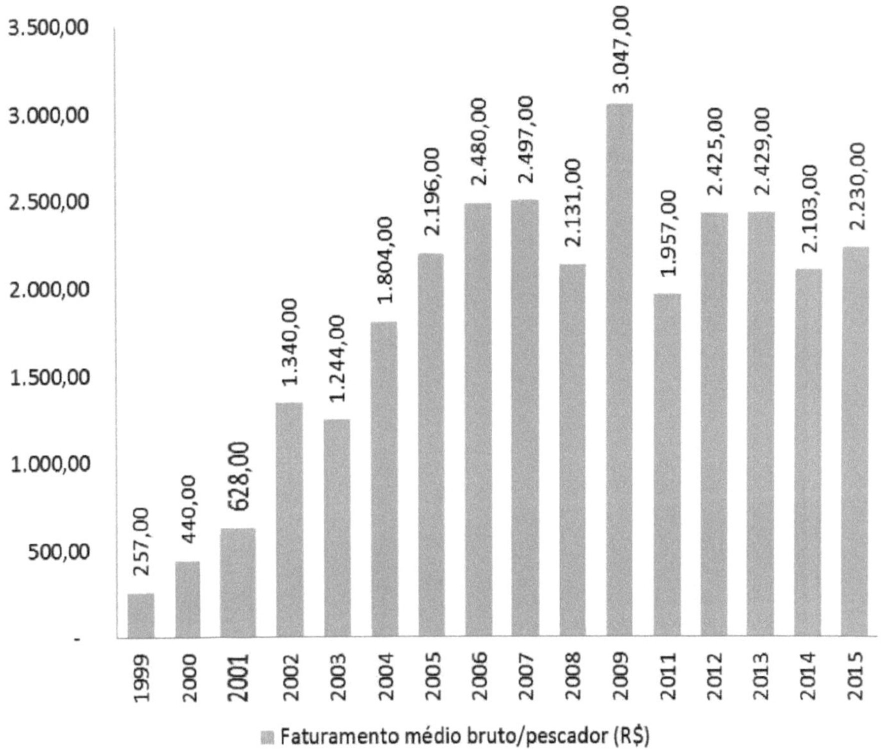

Figure 26: Average gross revenue/fisherman (R$) from pirarucu production.
Source: Mamirauá Institute.

The income from management has mainly been used to renovate and build the homes of the families involved, to buy various material goods (such as televisions, satellite dishes, fridges and small motorized boats) and to invest in the education of children and young people (LIMA, 2010).

Although pirarucu fishing is the main economic activity practiced by the riverside populations of the Mamirauá RDS, pirarucu management only generates income during the dry season, between October and November, when the lakes are isolated and the fish are more easily caught. During the rest of the months, the managers dedicate themselves to family farming, small livestock farming, community forest management and other activities (AMARAL, 2007).

Population growth of the species

The Jarauá Sector showed substantial increases in pirarucu densities in both categories (juveniles and adults) when compared to the first year of counting (Figure 27).

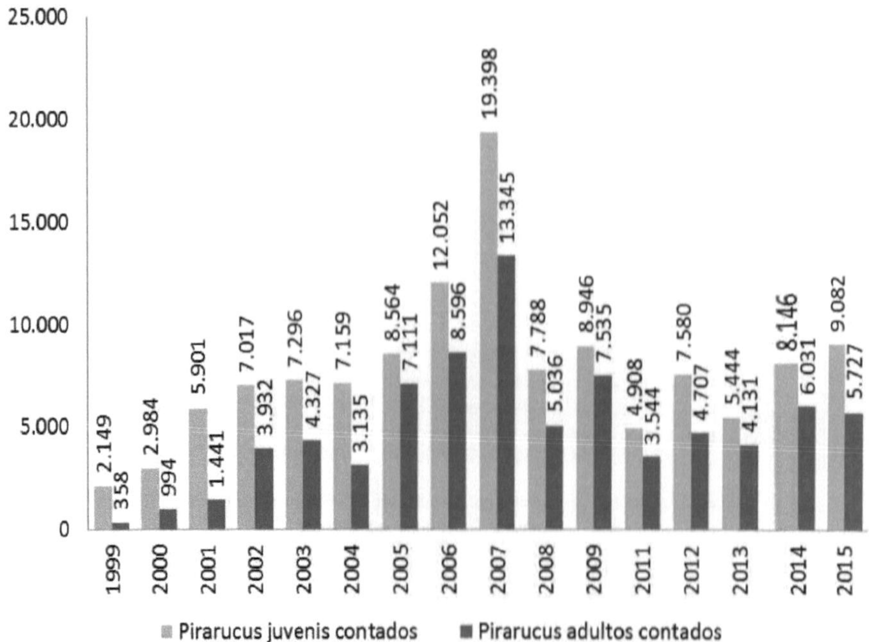

Figure 27: Evolution of juvenile and adult pirarucu counts in the Jarauá Sector.
Source: Mamirauá Institute.

When the abundance of pirarucu is analyzed by the sum of both categories, the Jarauá Sector shows general growth trends. In the Sector, the percentage increase in pirarucu densities from the first to the sixteenth year of counting was 490% (Figure 28).

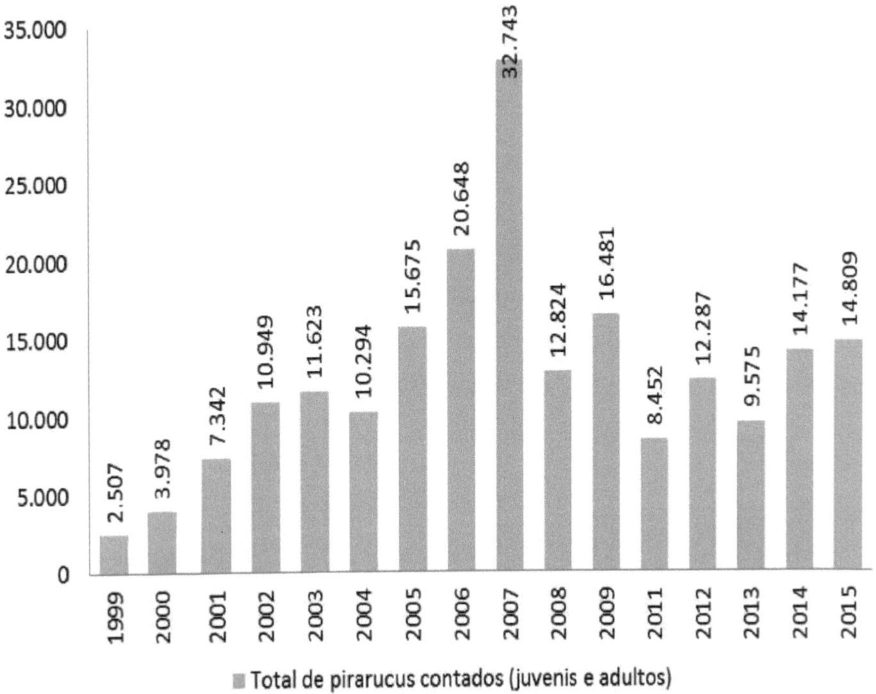

Figure 28: Population census of pirarucus counted (juveniles and adults) in the Sector.
Source: Mamirauá Institute.

It can be seen that in some years there were slight and strong declines in the abundance of these categories. Thus, in the years in which the counts indicated these population declines, reports made by fishermen and community members showed that one or more events occurred during the period in which the counts were carried out, such as the difficulties faced in carrying out the counts, variations in the water level and the unavailability of experienced fishermen to carry them out.

The fluctuations in pirarucu densities in Mamirauá's management regions may also have been influenced by illegal fishing, because despite legal efforts and monitoring, illegal fishing continues to be a problem for the development of pirarucu management systems in the Reserve (BESSA; LIMA, 2010; HRBEK; CROSSA; FARIAS, 2007; VIANA *et al.*, 2007).

These activities may have directly interfered with the results of the counts in the Jarauá Sector. The sharp drop in population from 2007 to 2008 in this location may be due, among other factors, to the fact that some lakes in the region were fished illegally, so, in addition to the fishing pressure itself, local fishermen reported that some fish had dispersed to other areas.

In addition, every year fishermen face unforeseen circumstances when carrying out the counts. This is mainly because in the Amazon there is a seasonality in fishing activity determined by the hydrological cycle, which promotes changes in the abundance and distribution of fish (Figures 29 and 30) (CASTRO; MCGRATH, 2010; COUTINHO, 2008; PETRERE *et al.*, 2007).

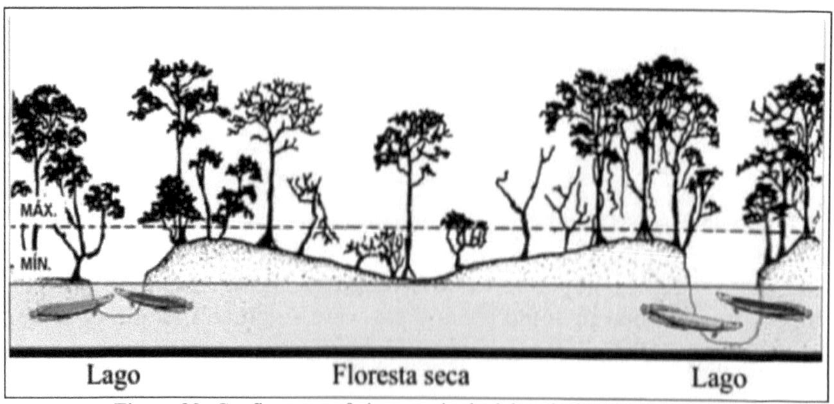

Figure 29: Confinement of pirarucus in the lakes during the drought.
Source: Araripe, 2008.

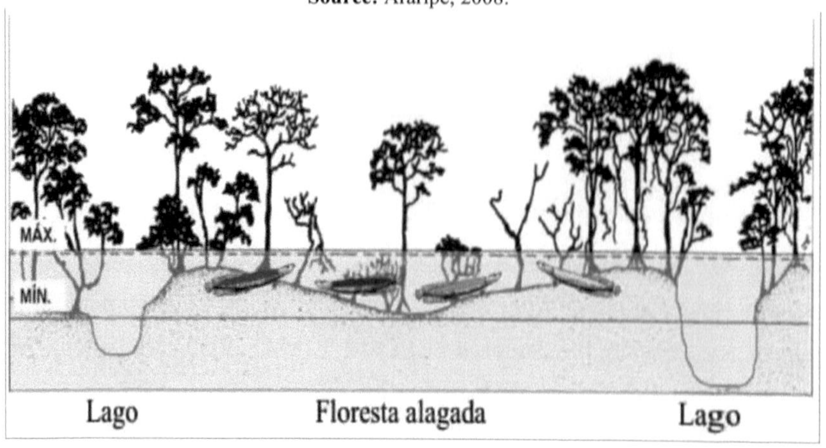

Figure 30: Dispersal of pirarucus from the lakes during the flood.
Source: Araripe, 2008.

Likewise, the counts are also directly influenced by the water level. Ideally, they should be carried out during the dry season, as soon as the lakes are isolated. This is because if there is no movement of pirarucu between the main channel and the lakes, there is less chance of the same individual being counted more than once.

On the other hand, water levels that are too low make it difficult or even impossible to access some lakes. In addition, very high water levels, even during the dry season, make it difficult for fishermen to observe pirarucu individuals. This is because the fish can disperse or take shelter in the vegetation. This fact was cited as occurring by fishermen in all the management regions of the Mamirauá RDS in 2004 (ARANTES; GARCEZ; CASTELLO, 2006).

Araripe *et al.* (2013) and Castello (2008) state that due to the dynamic environmental conditions and the impact on fishing, the pirarucu has developed an effective reproductive strategy (short term), longitudinal and lateral migrations (as a population compensation mechanism),
feeding strategy that changes according to its development and food availability, efficient patterns of area occupation (related to reproduction and feeding) and learning in response to fishing (ability to escape and/or reject fishing areas).

The status of pirarucu stocks in certain areas of the Mamirauá Reserve between 1993 and 1999 was assessed by Queiroz and Sardinha (1999). As in other areas of the Amazon basin, these authors also showed trends towards overfishing by recruitment and evidence of overfishing by growth of the species in this Reserve. They then predicted that production would tend to collapse and the population would be reduced by half by 2007 (VIANNA *et al.*, 2003, 2004).

Among the main challenges of the proposed management of the species in the Mamirauá RDS were the recovery of pirarucu stocks in their natural environments, which were once compromised, and the establishment of sustainable exploitation. The management proposal also aimed to generate more income for riverine families by increasing the value of the sustainably produced "pirarucu" product, as a way of compensating for the restrictive measures of the Reserve's management plan (VIANA *et al.*, 2003; VIANA *et al.,* 2007).

Viana *et al.* (2003) and Castello (2004) reported that the pirarucu population showed positive signs of stock recovery in the first few years of management, making it an innovative tool for conserving the resource. Fishermen's counts of pirarucu were also evaluated between 1999 and 2005 by Castello (2007) and Arantes (2009a) and showed that pirarucu population densities showed an upward trend in the locations monitored, increasing by around 5 times from the first to the last year studied.

After the implementation of management, positive changes were observed in aspects of pirarucu fishing in the Mamirauá Reserve (VIANA *et al.*, 2004, 2007; ARANTES; GARCEZ; CASTELLO, 2006; CASTELLO, 2007; AMARAL, 2007). Among other results, Arantes, Garcez and Castello (2006) and Viana *et al.* (2007), state that the average lengths of the catch have increased and the fishing of juveniles has reduced, in addition, the pirarucu population has shown signs of recovery. For example, in the locality where management was initially implemented, Setor Jarauá, the pirarucu population in 1999 was estimated through counts at 2,500 individuals and in 2006 at around 20,600.

In periods prior to the implementation of this fishing management system, between 1993 and 1998, the pirarucu population showed signs of overexploitation and was at a low population density (QUEIROZ; SARDINHA, 1999). During the period described by the authors, 80% of the catch was made up of young individuals (~120 cm in total length) and the average catch per unit effort (CPUE) was 2.41 kg/fisherman/hour. Since 1999, the first year of counting, the population density has grown substantially, the catch has been made up mostly of adult individuals and by 2003, the CPUE was already 17.25 kg/fisherman/hour (VIANA *et al.*, 2007).

Various community pirarucu management initiatives, based on the system developed in Mamirauá, are being implemented in other regions of the Amazon, such as the Baixo Juruá, Rio Jutaí and Auati-Paraná Extractive Reserves (RESEX's) in the state of Amazonas and in other regions of the states of Pará and Acre, as well as in countries such as Peru, Guyana and Colombia. Many of these systems have proved effective in conserving and recovering populations of the species.

Although most management strategies for fish species are based on size and age at sexual maturity, Lopes and Queiroz (2009), Castello, Stewart and Arantes (2013) recommend evaluating whether the growth and reproduction of the pirarucu population undergo changes as a result of increases in the population abundance of the species.

Araripe (2008) describes that the pirarucu's sexual maturity is slow, usually occurring at four or five years of age when they are 1.60 to 1.70 meters long and weigh around 40 to 50 kg. The author also states that the pirarucu has low fecundity despite the high fertility of the adults, noting that a female pirarucu is capable of producing around 11,000 fry per spawning, which is a low number when compared to other fish species.

Although a female pirarucu can generate a huge number of mature eggs (20,000 to 100,000), only a small proportion are fertilized by the male (around 430 eggs), giving rise to approximately 300 larvae, which in favorable environmental conditions will result in three or four adult fish each year (ARANTES *et al.*, 2011; COUTINHO, 2008; LOPES; QUEIROZ, 2009).

The minimum catch size for pirarucu in Brazil of 150 cm total length does not adequately protect the breeding stock, as the legislation authorizes the fishing of sexually immature females and this considerably increases the risk of the population collapsing. For example, around 20% of the individuals caught legally in Mamirauá between 1999 and 2007 measured between 150 and 157 cm in total length (AMARAL; QUEIROZ, 2011).

Increasing the minimum catch length of pirarucu in Mamirauá to 1.57 or 1.63 cm will contribute to their conservation, as the exploitation of overexploited pirarucu populations at low densities must be highly cautious, as the delay in sexual maturation seen at low densities will try to slow down population growth rates, which will further jeopardize the recovery of the population.

The counts made by the fishermen in this management system prove to be an important stage in the process. After nineteen years of developing the managed fishing model for pirarucu in the Mamirauá RDS, the counts carried out by the fishermen point to trends in the recovery of the species' densities in this Reserve. It should be emphasized that the issues that can make it difficult to carry out the counts, or even directly influence the quality of the information obtained from them, should be reassessed in order to guarantee the effectiveness of the models adopted.

In general, even with variations in pirarucu population densities over the years, the upward trends point to the current effectiveness of the fishing management model in contributing to the conservation of the species. This is partly because pirarucu management in Mamirauá is carried out in an adaptive and participatory manner (ARANTES; GARCEZ, 2006a; VIANA, 2003; 2004).

Considerations

The production of pirarucu managed in the Jarauá Sector, Mamirauá RSD, rose from 3,000 kg in 1999 to 67,422 kg in 2015. The number of fishermen benefiting from pirarucu management increased by 223% compared to the first year of management, so more families also benefited. The fishermen's income increased by around 767%, but despite this increase, the addition of new fishermen may mean a lower economic return. The population growth of pirarucu over 16 years has increased by around 490% in the Jarauá Sector areas, from 2,507 (2,149 juveniles and 358 adults) to 14,809 (9,082 juveniles and 5,727 adults).

The management of pirarucu in the Jarauá Sector of the Mamirauá RDS represents a sustainable technique from an economic, social and environmental point of view, and is contributing to the conservation of the species and the development of local populations.

REFERENCES

AMARAL, E. The community and the market: challenges in the commercialization of managed pirarucu from the Mamirauá and Amana reserves, Amazonas - Brazil. **Uakari**, v. 3, n. 2, p. 7-17, 2007.

AMARAL, E. **O Manejo Comunitário de pirarucu (*Arapaima Gigas*) como alternativa económica para os pescadores das Reservas Amaná e Mamirauá, Brasil**. 2009, 85 f. Dissertation (Master's Degree in Fisheries Management)
Natural Resources and Local Development) Federal University of Pará, Belém, 2009.

AMARAL, E.; QUEIROZ, H. Establishing sustainable quotas for pirarucu (*Arapaima gigas*) based on indicators other than counts. **Book of Abstracts of the VIII IDSM Annual Research Seminar** - Tefé: IDSM, 2011.

AMARAL, E. et al. **Management of pirarucus (*Arapaima gigas*) in floodplain lakes used exclusively by urban fishermen**. Tefé: IDSM, 2011.

AMARAL, E.; ALMEIDA, O. Productivity and economic efficiency of the pirarucu (*Arapaima gigas*) fishery in the management areas of the Amana and Mamirauá reserves. In: FIGUEIREDO, E. S. A. (Org.). **Biologia, conservado e manejo participativo de pirarucus na Pan-Amazónia,** Tefé: IDSM, 2013. p. 151-161.

ARANTES, C. C; GARCEZ, D. Densities of pirarucu (*Arapaima gigas*:

Teleostei, Osteoglossidae) in management regions in the Mamirauá and Amana Reserves, Amazonas, Brazil, **Uakari,** v.5, n. 1, p. 12-19, 2006a.

ARANTES, C. C.; GARCEZ, D. Avaliapäo das contagem de pirarucu (*Arapaima gigas*: Teleostei, Osteoglossidae) realizadas pelos pescadores das Reservas Mamirauá e Amana e implicapöes para o manejo, **Uakari,** v.2, n. 1, p. 32-49, 2006b.

ARANTES, C.C.; GARCEZ, D.S.; CASTELLO, L. Densities of pirarucu (*Arapaimagigas*, Teleostei, Osteoglossidae) in reserves of Mamirauá and Amana sustainable development, Amazonas, Brazil. **Uakari**, v. 2, n. 1, p. 37-43, 2006.

ARANTES, C. C.; CASTELLO, L.; GARCEZ, D. S. Variations between counts of *Arapaima gigas* (Schinz) (Osteogrlossomorpha, Osteoglossidae) made by individual fishermen in Mamirauá. **American Journal of Aquatic Sciences**, v. 2, n.1, p. 263-269, 2007.

ARANTES, C. C. **Evaluation of pirarucu counts by fishermen**. Tefé: IDSM, 2009a.

ARANTES, C. C. **Ecology of the pirarucu *Arapaima gigas* (Schinz, 1822) in the floodplain of the Mamirauá sustainable development reserve, Amazonas, Brazil**. 2009. 72 f. Dissertation (Master's in Zoology) - State University of Santa Cruz, Ilhéus, 2009b.

ARANTES, C.C. et al. Environmental influences on the distribution of arapaima in amazon floodplains. **Environmental Biology of Fishes**, v. 3, n. 1, p.1-11, 201 1.

ARAÚJO, L. M. S.; FREITAS, C. E. C. Use of ichthyofauna and other sources of protein by residents of traditional communities in the Amazon region. **Revista Agrogeoambiental**, v. 1, n. 2, p. 63-72, 2009.

ARARIPE J. **Genetics of pirarucus (*Arapaima gigas*) populations in the Mamirauá Reserve and considerations on the genetic structure for the species**. 2008, 116 f. Thesis (Doctorate in Zoology) - Federal University of Pará; Emilio Goeldi Museum of Pará, Belém, 2008.

ARARIPE, J. et al. Dispersal capacity and genetic structure of *Arapaima gigas* on different geographic scales using microsatellite markers. **PloS One**, v. 8, n. 2, p. 37-48, 2013.

BARTHEM, R. B.; GOULDING, M. **An Unexpected Ecosystem: The Amazon revealed by the fisheries**. Lima: Biblos, 2007.

BATISTA, V. S. et al. The state of fishing in the Amazon. In: BATISTA, V. S.; ISAAC, V. J. (Orgs.). **Peixes e pesca no Solimóes-Amazonas: uma avaliagao integrada**, 1ª ed. Manaus: IBAMA, 2012, p. 13-30.

BESSA, J. D. O.; LIMA, A.C. Fishing management of pirarucu (*Arapaima Gigas*) in the state of Amazonas: Mistakes, successes and future prospects. In: Seminário Internacional de Ciencias do Ambiente e Sustentabilidade na Amazónia, 1, 2010, Manaus. **Proceedings...** Amazonas: Federal University of Amazonas, 2010.

DEPIRARUCU LEATHER BAG . Available in: <http://www.taia.com.br/loja/moda-fashion/bolsa-em-couro-de-pirarucu-havana/>. Accessed on: October 10, 2017.

CASTELLO, L. A method to count pirarucu: fishers, assessment and management. **North American Journal of Fisheries Management**, v. 24, n. 7, p. 379-389, 2004.

CASTELLO, L. **A socio-ecological synthesis on the conservation of the pirarucu in floodplains of the Amazon**. New York: ProQuest, 2007.

CASTELLO, L. Lateral migration of *Arapaima gigas* in floodplains of the Amazon. **Ecology of Freshwater Fish**, v. 17, n. 4, p. 38 - 46, 2008

CASTELLO, L.; STEWART, D. J. Assessing CITES non-detriment findings procedures for arapaima in Brazil. **Journal of Applied Ichthyology**, v. 26, n. 3, p. 49-56, 2010.

CASTELLO, L.; STEWART, D. J.; ARANTES, C.C. Modeling population dynamics and conservation of arapaima in the Amazon. **Reviews in Fish Biology and Fisheries**, v. 21, n. 3, p. 623-640, 2011.

CASTELO, L.; STEWART, D.; ARANTES, C. C. What we know and need to do about pirarucu (*Arapaima* spp.) conservation in the Amazon. In: FIGUEIREDO, E. S. A. (Org.). **Biologia, conservaçao e manejo participativo de pirarucus na Pan-Amazônia,** Tefé: IDSM, 2013. p. 17-31.

CASTRO, F.; MCGRATH, D. Community-based lake management in the Amazon. **Strategic Partnerships**, v. 6, n. 12, p. 112-126, 2010.

COSTA, T.V. et al. Aspects of fish consumption and trade in Parintins. **Boletim do Instituto de Pesca**, Sao Paulo, v. 39, n. 1, 63-75, 2013.

PIRARUCU LEATHER. Available at: <http://peltica.com.br/#image-19>. Accessed on: October 10, 2017.

COUTINHO, E. S. S. **A dinámica populacional do pirarucu (*Arapaima gigas*) na Reserva de Desenvolvimento Sustentável Mamirauá (RDSM), Amazonia**. 2008, 191 f. Thesis (Doctorate in Science in Modeling) Computational) - Laboratorio Nacional de Ciencias Computacionais, Petrópolis, 2008.

PIRARUCU SCALE. Available at: <https://www.flickr.com/photos/iata/2117949183/>. Accessed on: October 10, 2017.

FERRARIS JR., C.J. Family Arapaimidae (Bonytongues). In: REIS, R.E.; KULLANDER, S.O.; FERRARIS JR, C.J. (Eds.). **Check list of the freshwater fishes of South and Central America**. Porto Alegre, Brazil: EDIPUCRS. p. 31,2003.

GARCEZ, D. S.; BOTERO, J.; FABRÉ, N. N. Characterization of subsistence and commercial fisheries practiced by riverine communities of floodplain areas in Manacapuru, Lower Solimoes, Amazonas, Brazil. **Tropical Journal of Fisheries and Aquatic Sciences**, v. 9, n. 1, p. 43-66, 2012.

GODINHO, H. P. et al. Gonadal morphology and reproductive traits of the Amazonian fish *Arapaima gigas* (Schinz, 1822). **Acta Zoologica**, v. 86, p. 289-294, 2005.

GONQALVES, A. C.T. Participatory management of pirarucu (*Arapaima gigas*) in the Mamirauá and Amana sustainable development reserves. In:

FIGUEIREDO, E. S. A. (Org.). **Biología, conservagao e manejo participativo de pirarucus na Pan-Amazonia,** Tefé: IDSM, 2013. p. 267277.

HRBEK, T. et al. Population genetic analysis of *Arapaima gigas*, one of the largest freshwater fishes of the Amazon basin: implications for its conservation. **Animal Conservation**, v. 8, n. 2, p. 297-308, 2005.

HRBEK, T.; CROSSA, M.; FARIAS, I. P. Conservation strategies for *Arapaima gigas* (Schinz, 1822) and the Amazonian várzea ecosystem. **Brazilian Journal of Biology**, v. 67, n. 4, p. 909-917, 2007.

MAMIRAUÁ SUSTAINABLE DEVELOPMENT INSTITUTE. **Balango Social**. Tefé: IDSM, 2014.

ISAAC, V. J. et al. Ecology of the fish fauna. In: BATISTA, V. S.; ISAAC, V. J. (Orgs.). **Peixes e pesca no Solimoes-Amazonas: uma avaliagao integrada**, 1ª ed. Manaus: IBAMA, 2012, p. 207-249.

JUNIOR, W.C.; ALMEIDA, O.T. Avaliagao do mercado da indústria pesqueira na Amazonia. In: ALMEIDA, O.T. (Org.). **The fishing industry in Amazonia**. Manaus: IBAMA, 2006, p. 122-169.

LIMA, D. Transformations in Mamirauá's domestic economy. **Uakari**, v. 6, n. 1, p. 9-26, 2010.

LIMA, L. G.; BATISTA, V. S. Ethnoichthyological studies on the pirarucu *Arapaima gigas* in Central Amazonia. **Acta Amazónica**, v. 42, n. 3, p. 337-344, 2012.

PIRARUCU LANGUAGE. Available at: <https://objetosdafloresta.com/2012/04/13/fva-e-mercado-de-manaus-124/>. Accessed on: October 10, 2017.

LOPES, K.; QUEIROZ, H. L. A review of the gonadal development stages of pirarucus *Arapaima gigas* (Schinz, 1822). **Uakari**, v. 5, n. 1, 3948, 2009.

LOPES, K.; QUEIROZ, H. L. Evaluation of the traditional knowledge of RDSM fishermen applied to the identification of the sex of pirarucus. **Uakari**, v. 5, n. 2, p. 59-66, 2010.

LOPES, K. et al. Notes on the reproductive biology of the pirarucu *Arapaima gigas* Schinz 1822. In: FIGUEIREDO, E. S. A. (Org.). **Biologia, conservapäo e manejo participativo de pirarucus na Pan-Amazónia,** Tefé: IDSM, 2013. p. 43-57.

MUELLER, O.; GREEN, A. ***Arapaima gigas*** **- Market Study: Current status of Arapaima global trade and perspectives on the Swiss, French and UK markets**. Geneva: UNCTAD, 2006.

PERALTA, N. et al. Household income and seasonality in communities of the Mamirauá RDS (1995-2005). **Uakari**, v. 5, n. 2, p. 7-19, 2009.

PETRERE JÚNIOR, M. et al. Amazonia: environments, resources and fishing. In: PETRERE JÚNIOR, M. (Org.). **O Sector pesqueiro na Amazonia: situapao atual e tendencias**, Manaus: IBAMA, 2007. p. 11-17.

QUEIROZ, H.L; SARDINHA, A.D. The preservation and sustainable use of pirarucus (*Arapaima gigas*, Osteoglossidae) in Mamirauá. In: QUEIROZ, H.L.; CRAMPTON, W.G.R. (Orgs.). **Strategies for managing fish resources in Mamirauá**. Brasilia: SCM; CNPq/MCT, 1999. p. 208-223.

QUEIROZ, H. L.; PERALTA, N. Sustainable Development Reserve: integrated management of natural resources and participatory management. In: GARAY, I.; BECKER, B. K. (Orgs.). **Human Dimensions of Biodiversity**. Petrópolis: Vozes, 2006. p. 447-476.

RUFFINO, M. L.; ISAAC, V. J. Las pesquerias del bajo Amazonas: problemas de manejo e desarcolo, **Acta Biologica**, v. 15, n. 2, p. 37-46, 1994.

RUFFINO, M. L. et al. Fisheries management in Amazonia and its monitoring. In: BATISTA, V. S.; ISAAC, V. J. (Orgs.). **Peixes e pesca no Solimoes- Amazonas: uma avaliapao integrada**, 1ª ed. Manaus: IBAMA, 2012, p. 251276.

RUFFINO, M. L. Status and trends of fishery resources of the Amazon basin in Brazil. In: WELCOMME, R. L.; VALBO-JORGENSEN, J.; HALLS, A. S. (Orgs.). **Inland fisheries evolution and management case studies from four continents**. 1st ed. Rome: FAO, 2014, p. 1-20.

SANTOS, G. M.; FERREIRA, E. J. G.; VAL, A. L. Fisheries resources and sustainability in the Amazon: facts and perspectives. **Hiléia**, v. 5, n. 8, p. 4377, 2010.

SANTOS, G.; FERREIRA, E. J. G.; ZUANON, J. A. S. **Commercial fish from Manaus**. Manaus: IBAMA, 2006.

SANTOS, A. F. **Journey through the Amazon River Basin and Colombia**. Sao Paulo: Baraúna, 2014.

EXOTIC PIRARUCU LEATHER SHOE. Available at: <https://www.silveradobotas.com.br/sapato-exotico-silverado-couro-de-pirarucu-cor-preto---100-leather/p>. Accessed on: October 10, 2017.

SILVA, R. B.; GONQALVES, A. C. T.; MARINHO, J. C. **Pirarucu Population Count and Census**. Tefé: IDSM, 2013.

STEWART, D. J. Re-description of *Arapaima agassizii* (Valenciennes), a rare fish from Brazil (Osteoglossomorpha, Osteoglossidae). **Copeia**, v. 2013, n. 1, p. 38-51,2013a.

STEWART, D. J. A new species of Arapaima (Osteoglossomorpha, Osteoglossidae) from the Solimoes River, Amazonas State, Brazil. **Copeia**, v. 2013, n. 3, p. 470-476, 2013b.

VIANA J. P. et al. Development of community fishing in the Mamirauá Sustainable Development Reserve. In: CAMPOS-ROZO C.; ULLOA, A. (Eds). **Fauna Socializada, Tendencias en el manejo participativo de la fauna en America Latina**. Bogotá: Colombian Institute of Anthropology and History, 2003. p. 335-351.

VIANA, J. P. et al. Economic Incentives for Sustainable Community Management of fishery Resourses in the Mamirauá Sustainable Development Reserve, Amazonas, Brazil. In: SILVIUS, K.M.; BODMER, R.; FRAGOSO, J.M.V. (Eds.). **People in Nature: wildlife conservation in South and Central America**. New York: Columbia University Press, 2004. p. 139-154.

VIANA, J. P. et al. Community management of pirarucu *Arapaima gigas* in the Mamirauá sustainable development reserve - Amazonas, Brazil. In: **Protected Aquatic Areas as a Fisheries Management Tool**, v. 4. Brasilia: IBAMA, 2007. p. 239-261.

Printed by Books on Demand GmbH, Norderstedt / Germany